# Naming and Picturing the Chemical Elements

*A Brief Look at the History of the names of each of the Chemical Elements in the Periodic Table*
Given in Three Charts:
Numerical Order
By Element number,
Alphabetical Order
By Chemical name
And
Alphabetical order
By symbol

Organized and Written by Catherine Jaime

# Introduction

Recently I was teaching a younger kids' class on scientists and inventors. That week's scientist was to be Dmitri Mendeleev (or "Mendeleyev," both are correct). It seemed a simple enough task. We would discuss his life briefly, take a look at the Periodic Table he helped create, play a game and be done with it. That's about how much time we would have and the introduction to Mendeleev and the Periodic Table would be done.

But, as so many of my plans seem to go, it didn't end up being quite that simple. When I dusted off my Periodic Table after decades of neglect, and started reminding myself of all the different elements displayed there, I took a closer look at their names. We had recently studied Marie Curie, so I wasn't surprised to find Polonium and Curium among the elements that I had never paid much attention to before. And Albert Einstein's namesake, Einsteinium, and Mendeleev and his element, Mendelevium came as no surprise. After our earlier studies this semester I was expecting those names. But where did some of these other names come from? Where did Argon or Zirconium come from? Or Fluorine or Hassium, or Cerium and Cobalt? It turned out they all had stories.

Here is a very brief look at each of those names and where they came from – one element at a time. You will probably notice fairly quickly that the names tended to come from one of four places: Early on the elements were generally named for *Greek or Latin words (those are shown in Italics);* then you start to see elements named for **specific locations (I've shown those in bold);** then the mythological names start to show up (they're underlined); and then, finally, some of the elements start getting named for specific individuals (those are highlighted).

I've included three versions of the charts – one with the elements listed numerically, according to their place on the Periodic Table, and the others with them listed alphabetically, to make it easier to look up specific elements – by their atomic number, name, or symbol.

Additionally, there is a matching game – one set of cards includes the element number, name, and symbol of one or more elements on each card – and the other with the brief history of the name(s). (There are 81 sets of cards in the game, since elements with common name features are included together on one card.)

I hope you and your students enjoy this unique look at the elements as much as we did!

Happy Learning!
*Cathy and Crew*

## Ideas for Using the Charts and the Game

I used these charts initially by asking my students to see if they could find the following elements:

*One that had been discovered in Europe (Europium, #63)*

*The one discovered in California (Californium, #98)*

*The one the Curies had named for Marie's home country of Poland.*
*(Polonium, #84)*

*The one named for Copernicus (Copernicium, #112)*

*The one named for Einstein (Einsteinium, #99)*

*The one name for Mendeleyev. (Mendelevium, #101)*

*The one named for Thor. (Thorium, #90)*

*The one named for the Titans (Titanium, #22)*

*Several named for planets.*
*(including Mercury, #80, Neptunium, #93, and Plutonium, #94)*

*One or more named for "rays"*
*(Radon, #86, Radium, #88, Actinium, #89)*

*Ones that have temporary names yet (#113, #115, #117, #118)*

You will likely find many other questions you could ask them. (Or let them ask each other!) My students were excited to have a copy of one of the charts to take home with them when the class was over.

For the game:
You can print pages 35 – 52 back to back if you would like to make flash cards.

Or you may print each page separately and use the cards as a Concentration game, where students match the card with the element name with the history of its name. (You can use one of the charts as an "answer key" for the game – or have the students use it to check each other.)

# Chemical Elements by Atomic Number

| # | Element/Symbol | History of the Name |
|---|---|---|
| 1 | Hydrogen **H** | Derived from the Greek words *"hydro"* and *"genes"* meaning "water-forming." |
| 2 | Helium **He** | From the Greek word *"helios"* meaning "sun." |
| 3 | Lithium **Li** | From the Greek word *"lithos"* meaning "stone." |
| 4 | Beryllium **Be** | From the Greek name for beryl, *"beryllo,"* meaning "precious blue-green color-of-sea-water stone." |
| 5 | Boron **B** | Most likely derived from the Arabic word *"buraq"* meaning "to glisten." |
| 6 | Carbon **C** | From the Latin word *"carbo"* meaning "charcoal." |
| 7 | Nitrogen **N** | Derived from the Greek words *"nitron"* and *"genes"* meaning "nitre-forming." |
| 8 | Oxygen **O** | Derived from the Greek *"oxy genes"* meaning "acid-forming." |
| 9 | Fluorine **F** | Derived from the Latin *"fluere"* meaning "to flow." |
| 10 | Neon **Ne** | Derived from the Greek *"neos"* meaning "new." |
| 11 | Sodium **Na** | Derived from the Arabic *"suda"* meaning "headache" – referring to the headache-alleviating properties of sodium carbonate. Symbol (Na) comes from Latin name *"natrium."* |
| 12 | Magnesium **Mg** | Named for **Magnesia, a district of Eastern Thessaly in Greece.** |
| 13 | Aluminium **Al** | Derived from the Latin name *"alumen"* meaning "bitter salt." |
| 14 | Silicon **Si** | Derived from the Latin word *"silex"* or *"silicis"* meaning "flint." |
| 15 | Phosphorus **P** | From the Greek word *"phosphoros"* meaning "bringer of light." |
| 16 | Sulfur **S** | Derived from the Sanskrit word *"sulvere"* or the Latin word *"sulfurium"* both meaning "to burn." |

| # | Element/Symbol | History of the Name |
|---|---|---|
| 17 | Chlorine **Cl** | Derived from the Greek word *"chloros"* meaning "greenish-yellow." |
| 18 | Argon **Ar** | Derived from the Greek word *"argos"* meaning "idle" or "inactive." |
| 19 | Potassium **K** | Derived from the Dutch word *"potaschen"* meaning "pot ashes." Symbol "K" comes from Latin *"kalium."* |
| 20 | Calcium **Ca** | From the Latin word *"calx"* meaning "lime." |
| 21 | Scandium **Sc** | Derived from **"Scandia," the Latin name for Scandinavia.** (The element was discovered in Sweden, a Scandinavian country.) |
| 22 | Titanium **Ti** | Derived from the <u>Titans, the sons of the Earth goddess of Greek mythology, Gaia, and Father Sky, Uranus.</u> |
| 23 | Vanadium **V** | Derived from <u>Vanadis, another name for the love goddess of Norse mythology, Freyja.</u> |
| 24 | Chromium **Cr** | Derived from the Greek word *"chroma"* meaning "color." |
| 25 | Manganese **Mn** | Possibly derived from the Latin word *"magnes"* meaning "magnet." |
| 26 | Iron **Fe** | Derived from the Anglo-Saxon name *"iren,"* but generally represented by the <u>symbol for Mars, the Roman God of War.</u> Symbol (Fe) came from the Latin name for iron – *"ferrum."* |
| 27 | Cobalt **Co** | Derived from the German word *"kobald"* meaning "goblin." |
| 28 | Nickel **Ni** | Derived from the German *"kupfernickel"* or the Swedish *"kopparnickel"* meaning either "St. Nicholas' copper" or "devil's copper." |
| 29 | Copper **Cu** | Derived from the Latin phrase *"Cyprium aes"* meaning a metal from **Cyprus**. Symbol (Cu) comes from the Latin name *"cuprum."* |
| 30 | Zinc **Zn** | Derived from the Persian word *"sing"* meaning "stone." |

| # | Element/Symbol | History of the Name |
|---|---|---|
| 31 | Gallium<br>**Ga** | From the Latin name for **France, "Gallia,"** after it was discovered by a French scientist. |
| 32 | Germanium<br>**Ge** | From the Latin name for **Germany, "Germania,"** in honor of the discovering scientist's homeland. |
| 33 | Arsenic<br>**As** | From the Greek name *"arsenikon."* |
| 34 | Selenium<br>**Se** | Derived from *"selene,"* the Greek name for the Moon. |
| 35 | Bromine<br>**Br** | From the Greek *"bromos"* meaning "stench." |
| 36 | Krypton<br>**Kr** | From the Greek *"kryptos"* meaning "hidden." |
| 37 | Rubidium<br>**Rb** | From the Latin *"rubidius"* meaning "deepest red." |
| 38 | Strontium<br>**Sr** | Named after **Strontian, a small town in Scotland** (where the element was discovered). |
| 39 | Yttrium<br>**Y** | Named for **Ytterby, Sweden,** where it was discovered. |
| 40 | Zirconium<br>**Zr** | From the Arabic word *"zargun,"* meaning "gold colored." |
| 41 | Niobium<br>**Nb** | Derived from <u>Niobe, the daughter of king Tantalus in Greek mythology.</u> |
| 42 | Molybdenum<br>**Mo** | From the Greek *"molybdos"* meaning "like lead." |
| 43 | Technetium<br>**Tc** | From the Greek "tekhnetos" meaning "artificial." |
| 44 | Ruthenium<br>**Ru** | From the Latin name for **Russia, "Ruthenia."** (One of the places the element was first discovered was Tartu, Russia.) |
| 45 | Rhodium<br>**Rh** | From the Greek *"rhodon"* or *"rhodos"* meaning "rose colored." |
| 46 | Palladium<br>**Pd** | Named after the asteroid <u>Pallas, which had been named after the Greek goddess of wisdom, Pallas.</u> |

| # | Element/Symbol | History of the Name |
|---|---|---|
| 47 | Silver **Ag** | From the Anglo-Saxon name *"siolfur."* Symbol (Ag) comes from the Latin word *"argentum."* |
| 48 | Cadmium **Cd** | From the Latin *"cadmia."* |
| 49 | Indium **In** | From the Latin *"indicium,"* meaning violet or indigo. |
| 50 | Tin **Sn** | From the Anglo-Saxon *"tin."* Symbol (Sn) comes from the Latin *"stannum".* |
| 51 | Antimony **Sb** | Derived from the Greek *"anti-monos"* meaning "not alone." Symbol (Sb) comes from Latin *"stibium."* |
| 52 | Tellurium **Te** | From the Latin *"tellus"* meaning "Earth." |
| 53 | Iodine **I** | From the Greek *"iodes"* meaning "violet." |
| 54 | Xenon **Xe** | From the Greek *"xenos"* meaning "stranger." |
| 55 | Caesium **Cs** | From the Latin *"caesius"* meaning "sky blue." |
| 56 | Barium **Ba** | From the Greek *"barys"* meaning "heavy." |
| 57 | Lanthanum **La** | From the Greek *"lanthanein"* meaning to "lie hidden." |
| 58 | Cerium **Ce** | Named for the asteroid, <u>Ceres, which had been named after the Roman goddess of agriculture.</u> |
| 59 | Praseodymium **Pr** | From the Greek *"prasios didymos"* meaning "green twin." |
| 60 | Neodymium **Nd** | From the Greek *"neos didymos"* meaning "new twin." |
| 61 | Promethium **Pm** | Named after <u>Prometheus, a Titan from Greek mythology who stole fire from Mount Olympus and gave it to mankind.</u> |
| 62 | Samarium **Sm** | From Samarskite, which had been named for Vasili Samarsky-Bykhovets, a Russian mining engineer. |

| # | Element/Symbol | History of the Name |
|---|---|---|
| 63 | Europium **Eu** | Named after **Europe.** (The element was discovered by two French scientists.) |
| 64 | Gadolinium **Gd** | Named in honor of Johan Gadolin, a Finnish chemist, mineralogist, and physicist. |
| 65 | Terbium **Tb** | Named for **Ytterby, Sweden,** where it was discovered. |
| 66 | Dysprosium **Dy** | From the Greek word *"dysprositos"* meaning "hard to get." |
| 67 | Holmium **Ho** | From the Latin name for **Stockholm, "Holmia."** |
| 68 | Erbium **Er** | Named for **Ytterby, Sweden,** where it was discovered. |
| 69 | Thulium **Tm** | Derived from **Thule, the ancient name for Scandinavia.** (Thulium was also discovered in Sweden, a Scandinavian country.) |
| 70 | Ytterbium **Yb** | Named for **Ytterby, Sweden,** where it was discovered. |
| 71 | Lutetium **Lu** | Derived from the Romans' name for **Paris, "Lutetia."** (Lutetium was discovered there.) |
| 72 | Hafnium **Hf** | Derived from the Latin name for **Copenhagen, "Hafnia"** (where it was discovered). |
| 73 | Tantalum **Ta** | Derived from the legendary Greek King Tantalus. |
| 74 | Tungsten **W** | From the Swedish *"tung sten"* meaning "heavy stone." Tungsten is also called "Wolfram" in some places, derived from the German "*wolf rahm*" meaning "wolf soot." Hence the symbol "W." |
| 75 | Rhenium **Re** | From the Latin name for the **Rhine, "Rhenus."** |
| 76 | Osmium **Os** | Derived from the Greek word *"osme"* meaning "smell." |
| 77 | Iridium **Ir** | Named for the Greek goddess of the rainbow, Iris. |
| 78 | Platinum **Pt** | From the Spanish *"platina"* meaning "little silver." |

| # | Element/Symbol | History of the Name |
|---|---|---|
| 79 | Gold **Au** | From the Latin word for gold - *"aurum."* |
| 80 | Mercury **Hg** | Named after the <u>planet, Mercury, which had been named for Mercury, the Roman messenger to the gods.</u> |
| 81 | Thallium **Tl** | From the Greek *"thallos"* meaning "a green twig." |
| 82 | Lead **Pb** | From the Anglo-Saxon word for the metal, *"lead."* The symbol (Pb) comes from the Latin word *"plumbum"* - since the Romans used lead in their plumbing. |
| 83 | Bismuth **Bi** | From the German word *"Bisemutum,"* from the words *"Weisse Masse"* meaning "white mass." |
| 84 | Polonium **Po** | From the Latin, *"Polonia,"* for **Poland,** the native country of Marie Curie, who first isolated the element with her husband, Pierre Curie. |
| 85 | Astatine **At** | From the Greek *"astatos"* meaning "unstable." |
| 86 | Radon **Rn** | First detected by the Curies, as it was emitted from Radium. Named for Radium, so also from the Latin for "ray." |
| 87 | Francium **Fr** | Named after **France,** where it was discovered. |
| 88 | Radium **Ra** | From the Latin *"radius"* meaning "ray." |
| 89 | Actinium **Ac** | From the Greek *"actinos"* meaning "a ray." |
| 90 | Thorium **Th** | Named after <u>Thor, the Scandinavian god of war.</u> |
| 91 | Protactinium **Pa** | From the Greek *"protos"* meaning "first," this element was named as "the parent of the element actinium." |
| 92 | Uranium **U** | Named after <u>the planet Uranus, which had been named after the Latinized version of the Greek god of the sky, Ouranos.</u> |

| # | Element/Symbol | History of the Name |
|---|---|---|
| 93 | Neptunium **Np** | Named after the planet Neptune, which had been named after the Roman god of the sea. |
| 94 | Plutonium **Pu** | Named after the then planet Pluto, which had been named after the Greek god of the underworld. |
| 95 | Americium **Am** | Named for **America,** since it was first made in the United States. |
| 96 | Curium **Cm** | Named in honor of Pierre and Marie Curie. |
| 97 | Berkelium **Bk** | Named after **Berkeley, California**, where it was first made. |
| 98 | Californium **Cf** | Named for the **University of California and state of California**, where the element was first made. |
| 99 | Einsteinium **Es** | Named after the renowned physicist Albert Einstein. |
| 100 | Fermium **Fm** | Named after the nuclear physicist Enrico Fermi. |
| 101 | Mendelevium **Md** | Named for Dmitri Mendeleev who produced one of the first periodic tables. |
| 102 | Nobelium **No** | Named for Alfred Nobel, the founder of the Nobel prize. |
| 103 | Lawrencium **Lr** | Named after Ernest O. Lawrence, the inventor of the cyclotron. |
| 104 | Rutherfordium **Rf** | Named in honor of New Zealand Chemist Ernest Rutherford. |
| 105 | Dubnium **Db** | Named for the **Russian town Dubna** where the element was first produced. |
| 106 | Seaborgium **Sg** | Named for Glenn T. Seaborg, an American chemist. |
| 107 | Bohrium **Bh** | Named for the Danish physicist Niels Bohr. |
| 108 | Hassium **Hs** | Derived from the **German state of Hesse** where Hassium was first produced. |
| 109 | Meitnerium **Mt** | Named for the Austrian physicist Lise Meitner. |

| # | Element/Symbol | History of the Name |
|---|---|---|
| 110 | Darmstadtium **Ds** | Named after **Darmstadt, Germany**, where the element was first produced. |
| 111 | Roentgenium **Rg** | Named in honor of the German physicist Wilhelm Conrad Röntgen. |
| 112 | Copernicium **Cn** | Named for the Renaissance mathematician and astronomer Nicolaus Copernicus. |
| 113 | (Ununtrium) **Uut** | Temporary Name, since discovered in 2015 |
| 114 | Flerovium **Fl** | Named after the Russian physicist Georgy Flerov (or Flyorov) who founded the institute where the element was discovered. |
| 115 | (Ununpentium) **Uup** | Temporary Name, since discovered in 2015 |
| 116 | Livermorium **Lv** | Named after the **Lawrence Livermore National Laboratory in Livermore, California.** |
| 117 | (Ununseptium) **Uus** | Temporary Name, since discovered in 2015 |
| 118 | (Ununoctium) **Uuo** | Temporary Name, since discovered in 2015 |

*Portrait of Dmitry Ivanovich Mendeleev*
*wearing the Edinburgh University professor robe.*
Ilya Repin, 1885

# *Chemical Elements*
# *Alphabetical by Name*

| # | Element/Symbol | History of the Name |
|---|---|---|
| 89 | Actinium **Ac** | From the Greek *"actinos"* meaning "a ray." |
| 13 | Aluminium **Al** | Derived from the Latin name *"alumen"* meaning "bitter salt." |
| 95 | Americium **Am** | Named for **America,** where it was first made. |
| 51 | Antimony **Sb** | Derived from the Greek *"anti-monos"* meaning "not alone." Symbol (Sb) comes from Latin *"stibium."* |
| 18 | Argon **Ar** | Derived from the Greek word *"argos"* meaning "idle" or "inactive." |
| 33 | Arsenic **As** | From the Greek name *"arsenikon."* |
| 85 | Astatine **At** | From the Greek *"astatos"* meaning "unstable." |
| 56 | Barium **Ba** | From the Greek *"barys"* meaning "heavy." |
| 97 | Berkelium **Bk** | Named after **Berkeley, California**, where it was first made. |
| 4 | Beryllium **Be** | From the Greek name for beryl, *"beryllo,"* meaning "precious blue-green color-of-sea-water stone." |
| 83 | Bismuth **Bi** | From the German word *"Bisemutum,"* from the words *"Weisse Masse"* meaning "white mass." |
| 107 | Bohrium **Bh** | Named for the Danish physicist Niels Bohr. |
| 5 | Boron **B** | Most likely derived from the Arabic word *"buraq"* meaning "to glisten." |
| 35 | Bromine **Br** | From the Greek *"bromos"* meaning "stench." |
| 48 | Cadmium **Cd** | From the Latin *"cadmia."* |
| 55 | Caesium **Cs** | From the Latin *"caesius"* meaning "sky blue." |
| 20 | Calcium **Ca** | From the Latin word *"calx"* meaning "lime." |

| # | Element/Symbol | History of the Name |
|---|---|---|
| 98 | Californium **Cf** | Named for the **University of California and state of California**, where the element was first made. |
| 6 | Carbon **C** | From the Latin word *"carbo"* meaning "charcoal." |
| 58 | Cerium **Ce** | Named for the asteroid, Ceres, which had been named after the Roman goddess of agriculture. |
| 17 | Chlorine **Cl** | Derived from the Greek word *"chloros"* meaning "greenish-yellow." |
| 24 | Chromium **Cr** | Derived from the Greek word *"chroma"* meaning "color." |
| 27 | Cobalt **Co** | Derived from the German word *"kobald"* meaning "goblin." |
| 112 | Copernicium **Cn** | Named for the Renaissance mathematician and astronomer Nicolaus Copernicus. |
| 29 | Copper **Cu** | Derived from the Latin phrase *"Cyprium aes"* meaning a metal from **Cyprus**. Symbol (Cu) comes from the Latin name *"cuprum."* |
| 96 | Curium **Cm** | Named in honor of Pierre and Marie Curie. |
| 110 | Darmstadtium **Ds** | Named after **Darmstadt, Germany**, where the element was first produced. |
| 105 | Dubnium **Db** | Named for the **Russian town Dubna** where the element was first produced. |
| 66 | Dysprosium **Dy** | From the Greek word *"dysprositos"* meaning "hard to get." |
| 99 | Einsteinium **Es** | Named after the renowned physicist Albert Einstein. |
| 68 | Erbium **Er** | Named for **Ytterby, Sweden,** where it was discovered. |
| 63 | Europium **Eu** | Named after **Europe.** (The element was discovered by two French scientists.) |
| 100 | Fermium **Fm** | Named after the nuclear physicist Enrico Fermi. |

| # | Element/Symbol | History of the Name |
|---|---|---|
| 114 | Flerovium<br>**Fl** | Named after the Russian physicist Georgy Flerov (or Flyorov) who founded the institute where the element was discovered. |
| 9 | Fluorine<br>**F** | Derived from the Latin *"fluere"* meaning "to flow." |
| 87 | Francium<br>**Fr** | Named after **France,** where it was discovered. |
| 64 | Gadolinium<br>**Gd** | Named in honor of Johan Gadolin, a Finnish chemist, mineralogist, and physicist. |
| 31 | Gallium<br>**Ga** | From the Latin name for **France, "Gallia,"** after it was discovered by a French scientist. |
| 32 | Germanium<br>**Ge** | From the Latin name for **Germany, "Germania,"** in honor of the discovering scientist's homeland. |
| 79 | Gold<br>**Au** | From the Latin word for gold - *"aurum."* |
| 72 | Hafnium<br>**Hf** | Derived from the Latin name for **Copenhagen, "Hafnia"** (where it was discovered). |
| 108 | Hassium<br>**Hs** | Derived from the **German state of Hesse** where Hassium was first produced. |
| 2 | Helium<br>**He** | From the Greek word *"helios"* meaning "sun." |
| 67 | Holmium<br>**Ho** | From the Latin name for **Stockholm, "Holmia."** |
| 1 | Hydrogen<br>**H** | Derived from the Greek words *"hydro"* and *"genes"* meaning "water-forming." |
| 49 | Indium<br>**In** | From the Latin *"indicium,"* meaning violet or indigo. |
| 53 | Iodine<br>**I** | From the Greek *"iodes"* meaning "violet." |
| 77 | Iridium<br>**Ir** | Named for the Greek goddess of the rainbow, Iris. |
| 26 | Iron<br>**Fe** | Derived from the Anglo-Saxon name *"iren,"* but generally represented by the symbol for Mars, the Roman God of War. Symbol (Fe) came from the Latin name for iron – *"ferrum."* |

| # | Element/Symbol | History of the Name |
|---|---|---|
| 36 | Krypton **Kr** | From the Greek *"kryptos"* meaning "hidden." |
| 57 | Lanthanum **La** | From the Greek *"lanthanein"* meaning to "lie hidden." |
| 103 | Lawrencium **Lr** | Named after Ernest O. Lawrence, the inventor of the cyclotron. |
| 82 | Lead **Pb** | From the Anglo-Saxon word for the metal, *"lead."* The symbol (Pb) comes from the Latin word *"plumbum"* - since the Romans used lead in their plumbing. |
| 3 | Lithium **Li** | From the Greek word *"lithos"* meaning "stone." |
| 116 | Livermorium **Lv** | Named after the **Lawrence Livermore National Laboratory in Livermore, California.** |
| 71 | Lutetium **Lu** | Derived from the Romans' name for **Paris, "Lutetia."** (Lutetium was discovered there.) |
| 12 | Magnesium **Mg** | Named for **Magnesia, a district of Eastern Thessaly in Greece.** |
| 25 | Manganese **Mn** | Possibly derived from the Latin word *"magnes"* meaning "magnet." |
| 109 | Meitnerium **Mt** | Named for the Austrian physicist Lise Meitner. |
| 101 | Mendelevium **Md** | Named for Dmitri Mendeleev who produced one of the first periodic tables. |
| 80 | Mercury **Hg** | Named after the planet, Mercury, which had been named for Mercury, the Roman messenger to the gods. |
| 42 | Molybdenum **Mo** | From the Greek *"molybdos"* meaning "like lead." |
| 60 | Neodymium **Nd** | From the Greek *"neos didymos"* meaning "new twin." |
| 10 | Neon **Ne** | Derived from the Greek *"neos"* meaning "new." |
| 93 | Neptunium **Np** | Named after the planet Neptune, which had been named after the Roman god of the sea. |

| # | Element/Symbol | History of the Name |
|---|---|---|
| 28 | Nickel **Ni** | Derived from the German *"kupfernickel"* or the Swedish *"kopparnickel"* meaning either "St. Nicholas' copper" or "devil's copper." |
| 41 | Niobium **Nb** | Derived from Niobe, the daughter of king Tantalus in Greek mythology. |
| 7 | Nitrogen **N** | Derived from the Greek words *"nitron"* and *"genes"* meaning "nitre-forming." |
| 102 | Nobelium **No** | Named for Alfred Nobel, the founder of the Nobel prize. |
| 76 | Osmium **Os** | Derived from the Greek word *"osme"* meaning "smell." |
| 8 | Oxygen **O** | Derived from the Greek *"oxy genes"* meaning "acid-forming." |
| 46 | Palladium **Pd** | Named after the asteroid Pallas, which had been named after the Greek goddess of wisdom, Pallas. |
| 15 | Phosphorus **P** | From the Greek word *"phosphoros"* meaning "bringer of light." |
| 78 | Platinum **Pt** | From the Spanish *"platina"* meaning "little silver." |
| 94 | Plutonium **Pu** | Named after the then planet Pluto, which had been named after the Greek god of the underworld. |
| 84 | Polonium **Po** | From the Latin, *"Polonia,"* for **Poland,** the native country of Marie Curie, who first isolated the element with her husband, Pierre Curie. |
| 19 | Potassium **K** | Derived from the Dutch word *"potaschen"* meaning "pot ashes." Symbol "K" comes from Latin *"kalium."* |
| 59 | Praseodymium **Pr** | From the Greek *"prasios didymos"* meaning "green twin." |
| 61 | Promethium **Pm** | Named after Prometheus, a Titan from Greek mythology who stole fire from Mount Olympus and gave it to mankind. |

| # | Element/Symbol | History of the Name |
|---|---|---|
| 91 | Protactinium **Pa** | From the Greek *"protos"* meaning "first," this element was named as "the parent of the element actinium." |
| 88 | Radium **Ra** | From the Latin *"radius"* meaning "ray." |
| 86 | Radon **Rn** | First detected by the Curies, as it was emitted from Radium. Named for Radium, so also from the Latin for "ray." |
| 75 | Rhenium **Re** | From the Latin name for the **Rhine, "Rhenus."** |
| 45 | Rhodium **Rh** | From the Greek *"rhodon"* or *"rhodos"* meaning "rose colored." |
| 111 | Roentgenium **Rg** | Named in honor of the German physicist Wilhelm Conrad Röntgen. |
| 37 | Rubidium **Rb** | From the Latin *"rubidius"* meaning "deepest red." |
| 44 | Ruthenium **Ru** | From the Latin name for **Russia, "Ruthenia."** (One of the places the element was first discovered was Tartu, Russia.) |
| 104 | Rutherfordium **Rf** | Named in honor of New Zealand Chemist Ernest Rutherford. |
| 62 | Samarium **Sm** | From Samarskite, which had been named for Vasili Samarsky-Bykhovets, a Russian mining engineer. |
| 21 | Scandium **Sc** | Derived from **"Scandia," the Latin name for Scandinavia.** (The element was discovered in Sweden, a Scandinavian country.) |
| 106 | Seaborgium **Sg** | Named for Glenn T. Seaborg, an American chemist. |
| 34 | Selenium **Se** | Derived from *"selene,"* the Greek name for the Moon. |
| 14 | Silicon **Si** | Derived from the Latin word *"silex"* or *"silicis"* meaning "flint." |
| 47 | Silver **Ag** | From the Anglo-Saxon name *"siolfur."* Symbol (Ag) comes from the Latin word *"argentum."* |

| # | Element/Symbol | History of the Name |
|---|---|---|
| 11 | Sodium<br>**Na** | Derived from the Arabic "*suda*" meaning "headache" – referring to the headache-alleviating properties of sodium carbonate. Symbol (Na) comes from Latin name "*natrium.*" |
| 38 | Strontium<br>**Sr** | Named after **Strontian, a small town in Scotland** (where the element was discovered). |
| 16 | Sulfur<br>**S** | Derived from the Sanskrit word "*sulvere*" or the Latin word "*sulfurium*" both meaning "to burn." |
| 73 | Tantalum<br>**Ta** | Derived from the legendary Greek King Tantalus. |
| 43 | Technetium<br>**Tc** | From the Greek "tekhnetos" meaning "artificial." |
| 52 | Tellurium<br>**Te** | From the Latin "*tellus*" meaning "Earth." |
| 65 | Terbium<br>**Tb** | Named for **Ytterby, Sweden,** where it was discovered. |
| 81 | Thallium<br>**Tl** | From the Greek "*thallos*" meaning "a green twig." |
| 90 | Thorium<br>**Th** | Named after Thor, the Scandinavian god of war. |
| 69 | Thulium<br>**Tm** | Derived from **Thule, the ancient name for Scandinavia.** (Thulium was also discovered in Sweden, a Scandinavian country.) |
| 50 | Tin<br>**Sn** | From the Anglo-Saxon "*tin.*" Symbol (Sn) comes from the Latin "*stannum*". |
| 22 | Titanium<br>**Ti** | Derived from the Titans, the sons of the Earth goddess of Greek mythology, Gaia, and Father Sky, Uranus. |
| 74 | Tungsten<br>**W** | From the Swedish "*tung sten*" meaning "heavy stone." Tungsten is also called "Wolfram" in some places, derived from the German "*wolf rahm*" meaning "wolf soot." Hence the symbol "W." |
| 92 | Uranium<br>**U** | Named after the planet Uranus, which had been named after the Latinized version of the Greek god of the sky, Ouranos. |

| #   | Element/Symbol      | History of the Name                                                                      |
| --- | ------------------- | ---------------------------------------------------------------------------------------- |
| 23  | Vanadium<br>**V**   | Derived from <u>Vanadis, another name for the love goddess of Norse mythology, Freyja.</u> |
| 54  | Xenon<br>**Xe**     | From the Greek *"xenos"* meaning "stranger."                                              |
| 70  | Ytterbium<br>**Yb** | Named for **Ytterby, Sweden,** where it was discovered.                                   |
| 39  | Yttrium<br>**Y**    | Named for **Ytterby, Sweden,** where it was discovered.                                   |
| 30  | Zinc<br>**Zn**      | Derived from the Persian word *"sing"* meaning "stone."                                   |
| 40  | Zirconium<br>**Zr** | From the Arabic word *"zargun,"* meaning "gold colored."                                  |
| 118 | (Ununoctium)<br>**Uuo**  | Temporary Name, since discovered in 2015                                             |
| 115 | (Ununpentium)<br>**Uup** | Temporary Name, since discovered in 2015                                             |
| 117 | (Ununseptium)<br>**Uus** | Temporary Name, since discovered in 2015                                             |
| 113 | (Ununtrium)<br>**Uut**   | Temporary Name, since discovered in 2015                                             |

# Periodic Table

Groups

| | 1 | 2 | 3 | 4 | 5 | 6 | 7 | 8 | 9 | 10 | 11 | 12 | 13 | 14 | 15 | 16 | 17 | 18 |
|---|---|---|---|---|---|---|---|---|---|---|---|---|---|---|---|---|---|---|
| **1** | 1 H hydrogen | | | | | | | | | | | | | | | | | 2 He helium |
| **2** | 3 Li lithium | 4 Be beryllium | | | | | | | | | | | 5 B boron | 6 C carbon | 7 N nitrogen | 8 O oxygen | 9 F fluorine | 10 Ne neon |
| **3** | 11 Na sodium | 12 Mg magnesium | | | | | | | | | | | 13 Al aluminum | 14 Si silicon | 15 P phosphorus | 16 S sulfur | 17 Cl chlorine | 18 Ar argon |
| **4** | 19 K potassium | 20 Ca calcium | 21 Sc scandium | 22 Ti titanium | 23 V vanadium | 24 Cr chromium | 25 Mn manganese | 26 Fe iron | 27 Co cobalt | 28 Ni nickel | 29 Cu copper | 30 Zn zinc | 31 Ga gallium | 32 Ge germanium | 33 As arsenic | 34 Se selenium | 35 Br bromine | 36 Kr krypton |
| **5** | 37 Rb rubidium | 38 Sr strontium | 39 Y yttrium | 40 Zr zirconium | 41 Nb niobium | 42 Mo molybdenum | 43 Tc technetium | 44 Ru ruthenium | 45 Rh rhodium | 46 Pd palladium | 47 Ag silver | 48 Cd cadmium | 49 In indium | 50 Sn tin | 51 Sb antimony | 52 Te tellurium | 53 I iodine | 54 Xe xenon |
| **6** | 55 Cs cesium | 56 Ba barium | 57-71 La-Lu lanthanides | 72 Hf hafnium | 73 Ta tantalum | 74 W wolfram | 75 Re rhenium | 76 Os osmium | 77 Ir iridium | 78 Pt platinum | 79 Au gold | 80 Hg mercury | 81 Tl thallium | 82 Pb plumbum | 83 Bi bismuth | 84 Po polonium | 85 At astatine | 86 Rn radon |
| **7** | 87 Fr francium | 88 Ra radium | 89-103 Ac-Lr actinides | 104 Rf rutherfordium | 105 Db dubnium | 106 Sg seaborgium | 107 Bh bohrium | 108 Hs hassium | 109 Mt meitnerium | 110 Ds darmstadtium | 111 Rg roentgenium | 112 Cn copernicium | 113 Uut ununtrium | 114 Fl flerovium | 115 Uup ununpentium | 116 Lv livermorium | 117 Uus ununseptium | 118 Uuo ununoctium |

Periods

| Lanthanides | 57 La lanthanum | 58 Ce cerium | 59 Pr praseodymium | 60 Nd neodymium | 61 Pm promethium | 62 Sm samarium | 63 Eu europium | 64 Gd gadolinium | 65 Tb terbium | 66 Dy dysprosium | 67 Ho holmium | 68 Er erbium | 69 Tm thulium | 70 Yb ytterbium | 71 Lu lutetium |
|---|---|---|---|---|---|---|---|---|---|---|---|---|---|---|---|
| Actinides | 89 Ac actinium | 90 Th thorium | 91 Pa protactinium | 92 U uranium | 93 Np neptunium | 94 Pu plutonium | 95 Am americium | 96 Cm curium | 97 Bk berkelium | 98 Cf californium | 99 Es einsteinium | 100 Fm fermium | 101 Md mendelevium | 102 No nobelium | 103 Lr lawrencium |

# Chemical Elements
# Alphabetical by Symbol

| # | Element/Symbol | History of the Name |
|---|---|---|
| 89 | **Ac** Actinium | From the Greek *"actinos"* meaning "a ray." |
| 47 | **Ag** Silver | From the Anglo-Saxon name *"siolfur."* Symbol (Ag) comes from the Latin word *"argentum."* |
| 13 | **Al** Aluminium | Derived from the Latin name *"alumen"* meaning "bitter salt." |
| 95 | **Am** Americium | Named for **America,** where it was first made. |
| 18 | **Ar** Argon | Derived from the Greek word *"argos"* meaning "idle" or "inactive." |
| 33 | **As** Arsenic | From the Greek name *"arsenikon."* |
| 85 | **At** Astatine | From the Greek *"astatos"* meaning "unstable." |
| 79 | **Au** Gold | From the Latin word for gold - *"aurum."* |
| 5 | **B** Boron | Most likely derived from the Arabic word *"buraq"* meaning "to glisten." |
| 56 | **Ba** Barium | From the Greek *"barys"* meaning "heavy." |
| 4 | **Be** Beryllium | From the Greek name for beryl, *"beryllo,"* meaning "precious blue-green color-of-sea-water stone." |
| 107 | **Bh** Bohrium | Named for the Danish physicist Niels Bohr. |
| 83 | **Bi** Bismuth | From the German word *"Bisemutum,"* from the words *"Weisse Masse"* meaning "white mass." |
| 97 | **Bk** Berkelium | Named after **Berkeley, California**, where it was first made. |
| 35 | **Br** Bromine | From the Greek *"bromos"* meaning "stench." |
| 6 | **C** Carbon | From the Latin word *"carbo"* meaning "charcoal." |
| 20 | **Ca** Calcium | From the Latin word *"calx"* meaning "lime." |

| # | Element/Symbol | History of the Name |
|---|---|---|
| 48 | **Cd** Cadmium | From the Latin *"cadmia."* |
| 58 | **Ce** Cerium | Named for the asteroid, <u>Ceres, which had been</u> <u>named after the Roman goddess of</u> <u>agriculture.</u> |
| 98 | **Cf** Californium | Named for the **University of California and state of California**, where the element was first made. |
| 17 | **Cl** Chlorine | Derived from the Greek word *"chloros"* meaning "greenish-yellow." |
| 96 | **Cm** Curium | Named in honor of Pierre and Marie Curie. |
| 112 | **Cn** Copernicium | Named for the Renaissance mathematician and astronomer Nicolaus Copernicus. |
| 27 | **Co** Cobalt | Derived from the German word *"kobald"* meaning "goblin." |
| 24 | **Cr** Chromium | Derived from the Greek word *"chroma"* meaning "color." |
| 55 | **Cs** Caesium | From the Latin *"caesius"* meaning "sky blue." |
| 29 | **Cu** Copper | Derived from the Latin phrase *"Cyprium aes"* meaning a metal from **Cyprus**. Symbol (Cu) comes from the Latin name *"cuprum."* |
| 105 | **Db** Dubnium | Named for the **Russian town Dubna** where the element was first produced. |
| 110 | **Ds** Darmstadtium | Named after **Darmstadt, Germany**, where the element was first produced. |
| 66 | **Dy** Dysprosium | From the Greek word *"dysprositos"* meaning "hard to get." |
| 68 | **Er** Erbium | Named for **Ytterby, Sweden,** where it was discovered. |
| 99 | **Es** Einsteinium | Named after the renowned physicist Albert Einstein. |
| 63 | **Eu** Europium | Named after **Europe.** (The element was discovered by two French scientists.) |

| # | Element/Symbol | History of the Name |
|---|---|---|
| 9 | **F** <br> Fluorine | Derived from the Latin *"fluere"* meaning "to flow." |
| 26 | **Fe** <br> Iron | Derived from the Anglo-Saxon name *"iren,"* but generally represented by the symbol for Mars, the Roman God of War. Symbol (Fe) came from the Latin name for iron – *"ferrum."* |
| 114 | **Fl** <br> Flerovium | Named after the Russian physicist Georgy Flerov (or Flyorov) who founded the institute where the element was discovered. |
| 100 | **Fm** <br> Fermium | Named after the nuclear physicist Enrico Fermi. |
| 87 | **Fr** <br> Francium | Named after **France,** where it was discovered. |
| 31 | **Ga** <br> Gallium | From the Latin name for **France, "Gallia,"** after it was discovered by a French scientist. |
| 64 | **Gd** <br> Gadolinium | Named in honor of Johan Gadolin, a Finnish chemist, mineralogist, and physicist. |
| 32 | **Ge** <br> Germanium | From the Latin name for **Germany, "Germania,"** in honor of the discovering scientist's homeland. |
| 1 | **H** <br> Hydrogen | Derived from the Greek words *"hydro"* and *"genes"* meaning "water-forming." |
| 2 | **He** <br> Helium | From the Greek word *"helios"* meaning "sun." |
| 72 | **Hf** <br> Hafnium | Derived from the Latin name for **Copenhagen, "Hafnia"** (where it was discovered). |
| 80 | **Hg** <br> Mercury | Named after the planet, Mercury, which had been named for Mercury, the Roman messenger to the gods. |
| 67 | **Ho** <br> Holmium | From the Latin name for **Stockholm, "Holmia."** |
| 108 | **Hs** <br> Hassium | Derived from the **German state of Hesse** where Hassium was first produced. |
| 53 | **I** <br> Iodine | From the Greek *"iodes"* meaning "violet." |

| # | Element/Symbol | History of the Name |
|---|---|---|
| 49 | **In**<br>Indium | From the Latin *"indicium,"* meaning violet or indigo. |
| 77 | **Ir**<br>Iridium | Named for the <u>Greek goddess of the rainbow, Iris.</u> |
| 19 | **K**<br>Potassium | Derived from the Dutch word *"potaschen"* meaning "pot ashes." Symbol "K" comes from Latin *"kalium."* |
| 36 | **Kr**<br>Krypton | From the Greek *"kryptos"* meaning "hidden." |
| 57 | **La**<br>Lanthanum | From the Greek *"lanthanein"* meaning to "lie hidden." |
| 3 | **Li**<br>Lithium | From the Greek word *"lithos"* meaning "stone." |
| 103 | **Lr**<br>Lawrencium | Named after Ernest O. Lawrence, the inventor of the cyclotron. |
| 71 | **Lu**<br>Lutetium | Derived from the Romans' name for **Paris, "Lutetia."** (Lutetium was discovered there.) |
| 116 | **Lv**<br>Livermorium | Named after the **Lawrence Livermore National Laboratory in Livermore, California.** |
| 101 | **Md**<br>Mendelevium | Named for Dmitri Mendeleev who produced one of the first periodic tables. |
| 12 | **Mg**<br>Magnesium | Named for **Magnesia, a district of Eastern Thessaly in Greece.** |
| 25 | **Mn**<br>Manganese | Possibly derived from the Latin word *"magnes"* meaning "magnet." |
| 42 | **Mo**<br>Molybdenum | From the Greek *"molybdos"* meaning "like lead." |
| 109 | **Mt**<br>Meitnerium | Named for the Austrian physicist Lise Meitner. |
| 7 | **N**<br>Nitrogen | Derived from the Greek words *"nitron"* and *"genes"* meaning "nitre-forming." |
| 11 | **Na**<br>Sodium | Derived from the Arabic *"suda"* meaning "headache" – referring to the headache-alleviating properties of sodium carbonate. Symbol (Na) comes from Latin name *"natrium."* |

| # | Element/Symbol | History of the Name |
|---|---|---|
| 41 | **Nb**<br>Niobium | Derived from Niobe, the daughter of king Tantalus in Greek mythology. |
| 60 | **Nd**<br>Neodymium | From the Greek *"neos didymos"* meaning "new twin." |
| 10 | **Ne**<br>Neon | Derived from the Greek *"neos"* meaning "new." |
| 28 | **Ni**<br>Nickel | Derived from the German *"kupfernickel"* or the Swedish *"kopparnickel"* meaning either "St. Nicholas' copper" or "devil's copper." |
| 102 | **No**<br>Nobelium | Named for Alfred Nobel, the founder of the Nobel prize. |
| 93 | **Np**<br>Neptunium | Named after the planet Neptune, which had been named after the Roman god of the sea. |
| 8 | **O**<br>Oxygen | Derived from the Greek *"oxy genes"* meaning "acid-forming." |
| 76 | **Os**<br>Osmium | Derived from the Greek word *"osme"* meaning "smell." |
| 15 | **P**<br>Phosphorus | From the Greek word *"phosphoros"* meaning "bringer of light." |
| 91 | **Pa**<br>Protactinium | From the Greek *"protos"* meaning "first," this element was named as "the parent of the element actinium." |
| 82 | **Pb**<br>Lead | From the Anglo-Saxon word for the metal, *"lead."* The symbol (Pb) comes from the Latin word *"plumbum"* - since the Romans used lead in their plumbing. |
| 46 | **Pd**<br>Palladium | Named after the asteroid Pallas, which had been named after the Greek goddess of wisdom, Pallas. |
| 61 | **Pm**<br>Promethium | Named after Prometheus, a Titan from Greek mythology who stole fire from Mount Olympus and gave it to mankind. |
| 84 | **Po**<br>Polonium | From the Latin, *"Polonia,"* for **Poland**, the native country of Marie Curie, who first isolated the element with her husband, Pierre Curie. |

| # | Element/Symbol | History of the Name |
|---|---|---|
| 59 | **Pr** Praseodymium | From the Greek *"prasios didymos"* meaning "green twin." |
| 78 | **Pt** Platinum | From the Spanish *"platina"* meaning "little silver." |
| 94 | **Pu** Plutonium | Named after the then planet Pluto, which had been named after the Greek god of the underworld. |
| 88 | **Ra** Radium | From the Latin *"radius"* meaning "ray." |
| 37 | **Rb** Rubidium | From the Latin *"rubidius"* meaning "deepest red." |
| 75 | **Re** Rhenium | From the Latin name for the **Rhine, "Rhenus."** |
| 104 | **Rf** Rutherfordium | Named in honor of New Zealand Chemist Ernest Rutherford. |
| 111 | **Rg** Roentgenium | Named in honor of the German physicist Wilhelm Conrad Röntgen. |
| 45 | **Rh** Rhodium | From the Greek *"rhodon"* or *"rhodos"* meaning "rose colored." |
| 86 | **Rn** Radon | First detected by the Curies, as it was emitted from Radium. Named for Radium, so also from the Latin for "ray." |
| 44 | **Ru** Ruthenium | From the Latin name for **Russia, "Ruthenia."** (One of the places the element was first discovered was Tartu, Russia.) |
| 16 | **S** Sulfur | Derived from the Sanskrit word *"sulvere"* or the Latin word *"sulfurium"* both meaning "to burn." |
| 51 | **Sb** Antimony | Derived from the Greek *"anti-monos"* meaning "not alone." Symbol (Sb) comes from Latin *"stibium."* |
| 21 | **Sc** Scandium | Derived from **"Scandia," the Latin name for Scandinavia.** (The element was discovered in Sweden, a Scandinavian country.) |
| 34 | **Se** Selenium | Derived from *"selene,"* the Greek name for the Moon. |

| # | Element/Symbol | History of the Name |
|---|---|---|
| 106 | **Sg** Seaborgium | Named for Glenn T. Seaborg, an American chemist. |
| 14 | **Si** Silicon | Derived from the Latin word *"silex"* or *"silicis"* meaning "flint." |
| 62 | **Sm** Samarium | From Samarskite, which had been named for Vasili Samarsky-Bykhovets, a Russian mining engineer. |
| 50 | **Sn** Tin | From the Anglo-Saxon *"tin."* Symbol (Sn) comes from the Latin *"stannum"*. |
| 38 | **Sr** Strontium | Named after **Strontian, a small town in Scotland** (where the element was discovered). |
| 73 | **Ta** Tantalum | Derived from the legendary Greek King Tantalus. |
| 65 | **Tb** Terbium | Named for **Ytterby, Sweden,** where it was discovered. |
| 43 | **Tc** Technetium | From the Greek "tekhnetos" meaning "artificial." |
| 52 | **Te** Tellurium | From the Latin *"tellus"* meaning "Earth." |
| 90 | **Th** Thorium | Named after Thor, the Scandinavian god of war. |
| 22 | **Ti** Titanium | Derived from the Titans, the sons of the Earth goddess of Greek mythology, Gaia, and Father Sky, Uranus. |
| 81 | **Tl** Thallium | From the Greek *"thallos"* meaning "a green twig." |
| 69 | **Tm** Thulium | Derived from **Thule, the ancient name for Scandinavia.** (Thulium was also discovered in Sweden, a Scandinavian country.) |
| 92 | **U** Uranium | Named after the planet Uranus, which had been named after the Latinized version of the Greek god of the sky, Ouranos. |
| 23 | **V** Vanadium | Derived from Vanadis, another name for the love goddess of Norse mythology, Freyja. |
| 74 | **W** | From the Swedish *"tung sten"* meaning "heavy |

| # | Element/Symbol | History of the Name |
|---|---|---|
| | Tungsten | stone." Also called "Wolfram" in some places, derived from the German "*wolf rahm*" meaning "wolf soot." Hence the symbol "W." |
| 54 | **Xe** Xenon | From the Greek *"xenos"* meaning "stranger." |
| 39 | **Y** Yttrium | Named for **Ytterby, Sweden,** where it was discovered. |
| 70 | **Yb** Ytterbium | Named for **Ytterby, Sweden,** where it was discovered. |
| 30 | **Zn** Zinc | Derived from the Persian word *"sing"* meaning "stone." |
| 40 | **Zr** Zirconium | From the Arabic word *"zargun,"* meaning "gold colored." |
| 118 | **Uuo** (Ununoctium) | Temporary Name, since discovered in 2015 |
| 115 | **Uup** (Ununpentium) | Temporary Name, since discovered in 2015 |
| 117 | **Uus** (Ununseptium) | Temporary Name, since discovered in 2015 |
| 113 | **Uut** (Ununtrium) | Temporary Name, since discovered in 2015 |

| | No. | | No. | | No. | | No. | | No. | | No. | | No. | | No. |
|---|---|---|---|---|---|---|---|---|---|---|---|---|---|---|---|
| H | 1 | F | 8 | Cl | 15 | Co & Ni | 22 | Br | 29 | Pd | 36 | I | 42 | Pt & Ir | 50 |
| Li | 2 | Na | 9 | K | 16 | Cu | 23 | Rb | 30 | Ag | 37 | Cs | 44 | Os | 51 |
| G | 3 | Mg | 10 | Ca | 17 | Zn | 24 | Sr | 31 | Cd | 38 | Ba & V | 45 | Hg | 52 |
| Bo | 4 | Al | 11 | Cr | 19 | Y | 25 | Ce & La | 33 | U | 40 | Ta | 46 | Tl | 53 |
| C | 5 | Si | 12 | Ti | 18 | In | 26 | Zr | 32 | Sn | 39 | W | 47 | Pb | 54 |
| N | 6 | P | 13 | Mn | 20 | As | 27 | Di & Mo | 34 | Sb | 41 | Nb | 48 | Bi | 55 |
| O | 7 | S | 14 | Fe | 21 | Se | 28 | Ro & Ru | 35 | Te | 43 | Au | 49 | Th | 56 |

*John Newlands' Periodic Table from 1866.*

# Element Game Cards

| Reihen | Gruppe I. — $R^2O$ | Gruppe II. — RO | Gruppe III. — $R^2O^3$ | Gruppe IV. $RH^4$ $RO^2$ | Gruppe V. $RH^3$ $R^2O^5$ | Gruppe VI. $RH^2$ $RO^3$ | Gruppe VII. RH $R^2O^7$ | Gruppe VIII. — $RO^4$ |
|---|---|---|---|---|---|---|---|---|
| 1 | H=1 | | | | | | | |
| 2 | Li=7 | Be=9.4 | B=11 | C=12 | N=14 | O=16 | F=19 | |
| 3 | Na=23 | Mg=24 | Al=27.3 | Si=28 | P=31 | S=32 | Cl=35.5 | |
| 4 | K=39 | Ca=40 | —=44 | Ti=48 | V=51 | Cr=52 | Mn=55 | Fe=56, Co=59, Ni=59, Cu=63. |
| 5 | (Cu=63) | Zn=65 | —=68 | —=72 | As=75 | Se=78 | Br=80 | |
| 6 | Rb=85 | Sr=87 | ?Yt=88 | Zr=90 | Nb=94 | Mo=96 | —=100 | Ru=104, Rh=104, Pd=106, Ag=108. |
| 7 | (Ag=108) | Cd=112 | In=113 | Sn=118 | Sb=122 | Te=125 | J=127 | |
| 8 | Cs=133 | Ba=137 | ?Di=138 | ?Ce=140 | — | — | — | — — — — |
| 9 | (—) | — | — | — | — | — | — | |
| 10 | — | — | ?Er=178 | ?La=180 | Ta=182 | W=184 | — | Os=195, Ir=197, Pt=198, Au=199. |
| 11 | (Au=199) | Hg=200 | Tl=204 | Pb=207 | Bi=208 | — | — | |
| 12 | — | — | — | Th=231 | — | U=240 | — | — — — — |

*Dmitri Mendeleev's 1871 Periodic Table*
*with room for undiscovered elements.*

| | | |
|---|---|---|
| #1<br>Hydrogen<br>H | #2<br>Helium<br>He | #3<br>Lithium<br>Li |
| #4<br>Beryllium<br>Be | #5<br>Boron<br>B | #6<br>Carbon<br>C |
| #7<br>Nitrogen<br>N | #8<br>Oxygen<br>O | #9<br>Fluorine<br>F |

| | | |
|---|---|---|
| From the Greek word "lithos" meaning "stone." | From the Greek word "helios" meaning "sun." | Derived from the Greek words "hydro" and "genes" meaning "water-forming." |
| From the Latin word "carbo" meaning "charcoal." | Most likely derived from the Arabic word "buraq" meaning "to glisten." | From the Greek name meaning "precious blue-green color-of-sea-water stone." |
| Derived from the Latin "fluere" meaning "to flow." | Derived from the Greek "oxy genes" meaning "acid-forming." | Derived from the Greek words "nitron" and "genes" meaning "nitre-forming." |

| | | |
|---|---|---|
| #10<br>Neon<br>N | #11<br>Sodium<br>Na | #12<br>Magnesium<br>Mg |
| #13<br>Aluminium<br>Al | #14<br>Silicon<br>Si | #15<br>Phosphorus<br>P |
| #16<br>Sulfur<br>S | #17<br>Chlorine<br>Cl | #18<br>Argon<br>Ar |

| | | |
|---|---|---|
| Named for Magnesia, a district of Eastern Thessaly in Greece. | Derived from the Arabic "suda" meaning "headache." | Derived from the Greek "neos" meaning "new." |
| From the Greek word "phosphoros" meaning "bringer of light." | Derived from the Latin word "silex" or "silicis" meaning "flint." | Derived from the Latin name "alumen" meaning "bitter salt." |
| Derived from the Greek word "argos" meaning "idle" or "inactive." | Derived from the Greek word "chloros" meaning "greenish-yellow." | Derived from the word "sulvere" or the word "sulfurium" both meaning "to burn." |

| | | |
|---|---|---|
| #19<br>Potassium<br>K | #20<br>Calcium<br>Ca | #21<br>Scandium,<br>Sc<br>#69<br>Thulium,<br>Tm |
| #22<br>Titanium<br>Ti | #23<br>Vanadium<br>V | #24<br>Chromium<br>Cr |
| #25<br>Maganese<br>Mn | #26<br>Iron<br>Fe | #27<br>Cobalt<br>Co |

| | | |
|---|---|---|
| Derived from names for Scandinavia. | From the Latin word "calx" meaning "lime." | Derived from the Dutch word "potaschen" meaning "pot ashes." |
| Derived from the Greek word "chroma" meaning "color." | Derived from Vanadis, the love goddess of Norse mythology, Freyja. | Derived from the Titans. |
| Derived from the German word "kobald" meaning "goblin." | Derived from the Anglo-Saxon name "iren." | Possibly derived from the Latin word "magnes" meaning "magnet." |

| #28 Nickel Ni | #29 Copper Cu | #30 Zinc Zn |
|---|---|---|
| #31 Gallium, Ga #71 Lutetium, Lu #87 Francium, Fr | #32 Germanium, Ge #108 Hassium, Hs #110 Darmstadtium, Ds | #33 Arsenic As |
| #34 Selenium Se | #35 Bromine Br | #36 Krypton, Kr #57 Lanthanum, La |

| | | |
|---|---|---|
| Derived from the Persian word "sing" meaning "stone." | Derived from the Latin phrase "Cyprium aes" meaning a metal from Cyprus. | Derived from "kupfernickel" or "kopparnickel" meaning "St. Nicholas' copper" or "devil's copper." |
| From the Greek name "arsenikon." | Named for Germany and places in Germany. | From various names for France and Paris. |
| From the Greek or Latin names for "hidden" or "to lie hidden." | From the Greek "bromos" meaning "stench." | Derived from "selene," the Greek name for the Moon. |

| #37 Rubidium Rb | #38 Strontium Sr | #39 Yttrium, Y #65 Terbium, Tb #68 Erbium, Er #70 Ytterbium, Yb |
|---|---|---|
| #40 Zirconium Zr | #41 Niobium Nb | #42 Molybdenum Mo |
| #43 Technetium Tc | #44 Ruthenium, Ru #105 Dubnium, Db | #45 Rhodium Rh |

| | | |
|---|---|---|
| Named for Ytterby, Sweden, where they were discovered. | Named after Strontian, Scotland (where the element was discovered). | From the Latin "rubidius" meaning "deepest red." |
| From the Greek "molybdos" meaning "like lead." | Derived from Niobe, the daughter of king Tantalus in Greek mythology. | From the Arabic word "zargun," meaning "gold colored." |
| From the Greek "rhodon" or "rhodos" meaning "rose colored." | Named for Russia and a town in Russia. | From the Greek "tekhnetos" meaning "artificial." |

| #46 Palladium, Pd<br>#52 Tellurium, Te<br>#58 Cerium, Ce<br>#80 Mercury, Hg<br>#92 Uranium, U<br>#93 Neptunium, Np<br>#94 Plutonium, Pu | #47<br>Silver<br>Ag | #48<br>Cadmium<br>Cd |
|---|---|---|
| #49<br>Indium<br>In | #50<br>Tin<br>Sn | #51<br>Antimony<br>Sb |
| #53<br>Iodine<br>I | #54<br>Xenon<br>X | #55<br>Caesium<br>Cs |

| From the Latin "cadmia." | From the Anglo-Saxon name "siolfur." | Named for a variety of asteroids and planets. |
|---|---|---|
| Derived from the Greek "anti-monos" meaning "not alone." | From the Anglo-Saxon "tin." | From the Latin "indicium," meaning violet or indigo. |
| From the Latin "caesius" meaning "sky blue." | From the Greek "xenos" meaning "stranger." | From the Greek "iodes" meaning "violet." |

| | | |
|---|---|---|
| **#56**<br>Barium, Ba<br><br>**#74**<br>Tungsten, W | **#59**<br>Praseodymium<br>**Pr** | **#60**<br>Neodymium<br>**Nd** |
| **#61**<br>Promethium<br>**Pm** | **#62**<br>Samarium<br>**Sm** | **#63**<br>Europium<br>**Eu** |
| #64 Gadolinium, Gd<br>#96 Curium, Cm<br>#99 Einsteinium, Es<br>#100 Fermium, Fm<br>#101 Mendelevium, Md<br>#102 Nobelium, No<br>#103 Lawrencium, Lr | **#66**<br>Dysprosium<br>**Dy** | **#67**<br>Holmium<br>**Ho** |

| | | |
|---|---|---|
| From the Greek "neos didymos" meaning "new twin." | From the Greek "prasios didymos" meaning "green twin." | From the Greek "barys" for "heavy" or the Swedish "tung sten" meaning "heavy stone." |
| Named after Europe. | From Samarskite. | Named after Prometheus, a Titan from Greek mythology. |
| From the Latin name for Stockholm, "Holmia." | From the Greek word "dysprositos" meaning "hard to get." | Named in honor of a variety of Scientists. |

| #72 Hafnium Hf | #73 Tantalum Ta | #75 Rhenium Re |
|---|---|---|
| #76 Osmium Os | #77 Iridium Ir | #78 Platinum Pt |
| #79 Gold Au | #81 Thallium Tl | #82 Lead Pb |

| | | |
|---|---|---|
| From the Latin name for the Rhine, "Rhenus." | Derived from the legendary Greek King Tantalus. | Derived from the Latin name for Copenhagen, "Hafnia." |
| From the Spanish "platina" "little silver." | Named for the Greek goddess of the rainbow, Iris. | Derived from the Greek word "osme" meaning "smell." meaning |
| From the Anglo-Saxon word for the metal, "lead." | From the Greek "thallos" meaning "a green twig." | From the Latin word for gold - "aurum." |

| #83<br>Bismuth<br>Bi | #84<br>Polonium<br>Po | #85<br>Astatine<br>At |
|---|---|---|
| #86<br>Radon, Rn<br>#88<br>Radium, Ra<br>#89<br>Actinium, Ac | #90<br>Thorium<br>Th | #91<br>Protactinium<br>Pa |
| #95<br>Americium, Am<br>#97<br>Berkelium, Bk<br>#98<br>Californium, Cf<br>#116<br>Livermorium, Lv | #104 Rutherfordium Rf<br>#106 Seaborgium Sg<br>#107 Bohrium Bh<br>#109 Meitnerium Mt<br>#111 Roentgenium Rg<br>#112 Copernicium Cn<br>#114 Flerovium Fl | #113<br>Ununtrium Uut<br>#115<br>Ununpentium Uup<br>#117<br>Ununseptium Uus<br>#118<br>Ununoctium Uuo |

| | | |
|---|---|---|
| From the Greek "astatos" meaning "unstable." | From the Latin, "Polonia," for Poland. | From the words "Weisse Masse" meaning "white mass." |
| From the Greek "protos" meaning "first." | Named after Thor, the Scandinavian god of war. | From the Greek or Latin words meaning "ray." |
| Recently discovered; still waiting for permanent names. | Named for a variety of Scientists. | Named for a variety of places in the United States, particularly California. |

# Picturing the Chemical Elements

## 100 ELEMENT PICTURE CARDS REPRESENTING EACH OF THE ELEMENTS WITH A RELATED PICTURE, INCLUDING: MAPS, STAMPS, COAT OF ARMS, SCIENTISTS

# Introduction

This project could be used by itself, or as a complement to "Naming the Chemical Elements." Suggestions for using the cards:

Students can arrange them in order by:
Their atomic number
Alphabetically by Element Name or Symbol
Chronologically by the dates on the cards
The occupations of the scientists who discovered them, or for whom they were named.
Geographically by the countries mentioned on the cards

Two sets of card can be printed and a matching game can be played (though you may want to do that with just a subset of the cards at a time, so as not to overwhelm those who are playing!)

I'm sure you and your students will come up with other fun ways to enjoy the cards.

There are also three different versions of the Periodic Table included, so that students can see that the Periodic Table hasn't always been designed the way it currently looks.

Happy Learning
*Cathy and Crew*

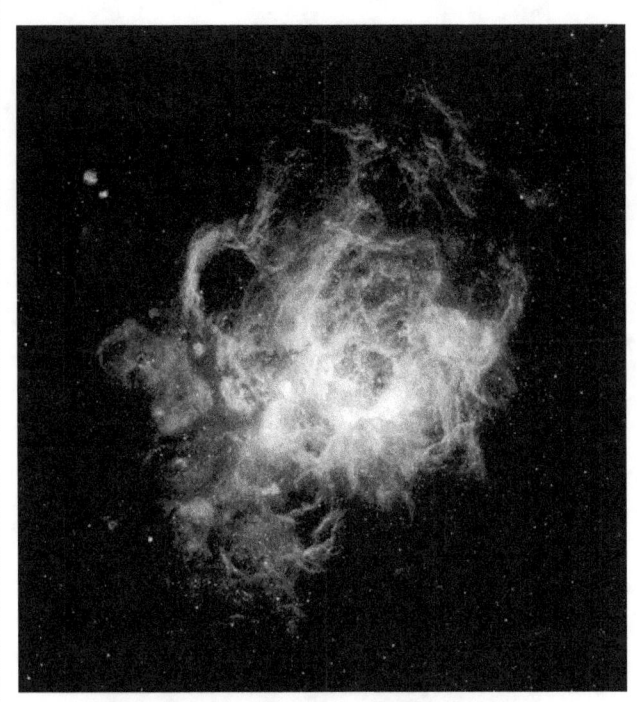

*NGC 604, ionized Hydrogen
in the Triangulum Galaxy.
Hydrogen was discovered in 1766.*
**#1, Hydrogen, H**

*Vanity Fair's Caricature of
Sir William Ramsay, Scottish chemist
who discovered the Noble Gases
and received the Nobel Prize
in Chemistry in 1904.*

**#2, Helium, He   #10, Neon, N**
**#18, Argon, Ar   #36, Krypton, Kr**
**#54, Xenon, X**

*Johan August Arfwedson,
Swedish chemist, is credited
with the discovery
of Lithium in 1817.*
**#3, Lithium, Li**

*Louis-Nicolas Vauquelin,
French chemist,
analyzed Beryllium in 1798.*
**#4, Beryllium, Be**

*Discovered in France in 1808.*
**#5, Boron, B**

*Antoine-Laurent de Lavoisier, French chemist, recognized Carbon in 1789.*
**#6, Carbon, C**

*Isolated in Scotland in 1772.*
**#7, Nitrogen, N**

*Robert H. Goddard (American) with a liquid oxygen-gasoline rocket in 1926.*
**#8, Oxygen, O**

*1556 woodcut illustration - showing fluorite being used as an additive to lower the melting point of metals during smelting.*
**#9, Fluorine, F**

*Close Up of Sodium Silicate. Isolated by Cornish chemist in 1807.*
**#11, Sodium, S**

*Joseph Black, Scottish chemist who discovered Magnesium in 1755.*
**#12, Magnesium, Mg**

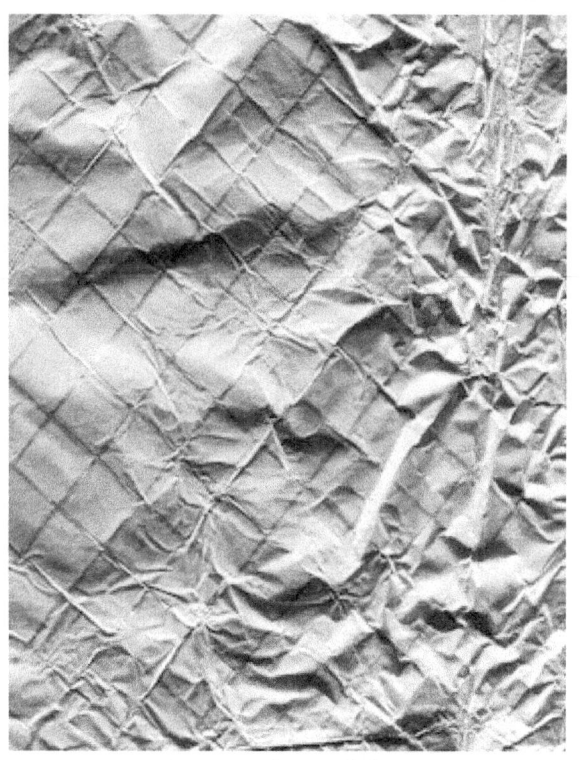

*First isolated by a Danish chemist in 1825.*
**#13, Aluminium, Al**

*Predicted by de Lavoisier in 1787.*
**#14, Silicon, Si**

*The Alchemist Discovering
Phosphorus
Painting by Joseph Wright in 1771.*
**#15, Phosphorus, P**

*Discovered by the Chinese
more than 4,000 years ago.*
**#16, Sulfur, S**

*Discovered by a
Swedish chemist in 1774.*
**#17, Chlorine, Cl**

*One of many elements
discovered by Cornish chemist,
Humphry Davy, this one in 1807.*
**#19, Potassium, K**

*From the Latin word meaning
"lime."
Humphry Davy discovered
Calcium in 1808.*
**#20, Calcium, Ca**

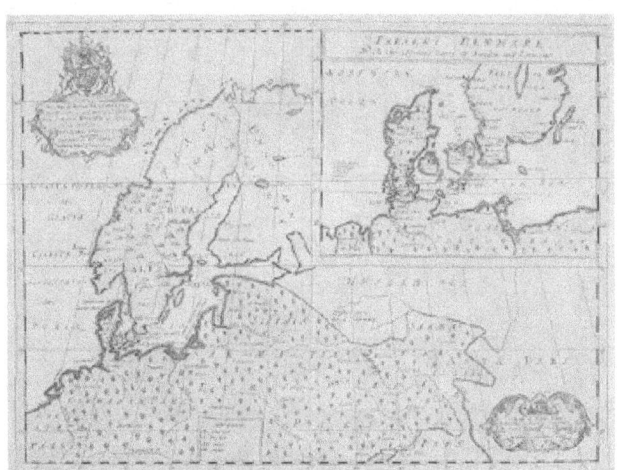

*Named for Scandinavia
after it was discovered
by a Swedish chemist in 1879.*
**#21, Scandium, Sc**

*Discovered by a
British mineralogist in 1791.*
**#22, Titanium, Ti**

*Discovered by Andrés Manuel del Río, a Spanish–Mexican scientist and naturalist, in 1801.*
**#23, Vanadium, V**

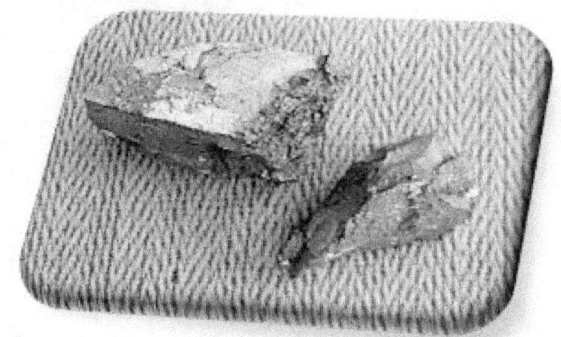

*Discovered by a French chemist in 1797.*
**#24, Chromium, Cr**

*Johan Gottlieb Gahn is given credit for first isolating pure Manganese in 1774.*
**#25, Maganese, Mn**

*An Iron Forge*
*Painting by Joseph Wright in 1772.*
**#26, Iron, Fe**

*Cobalt Blue Glassware
Cobalt has been used to color
glass for at least 3500 years.*
**#27, Cobalt, Co**

*Ravensthorpe Nickel
Operation Plant in Australia
where Nickel has been
mined since 2002.*
**#28, Nickel, Ni**

*Bingham Copper Mine in Utah where
Copper has been mined since 1863.*
**#29, Copper, Cu**

*Andreas Sigismund Marggraf
is given credit for first
isolating pure Zinc in 1746.*
**#30, Zinc, Zn**

*French chemist François Lecoq de Boisbaudran discovered Gallium, Samarium, and Dysprosium.*
**#31, Gallium, Ga** (1875)
**#62, Samarium, Sm** (1879)
**#66, Dysprosium, Dy** (1886)

*Named for Germany after discovered by a German chemist in 1886.*
**#32, Germanium, Ge**

*Satirical cartoon by Honoré Daumier of a chemist giving a public demonstration of Arsenic in 1841.*
**#33, Arsenic, As**

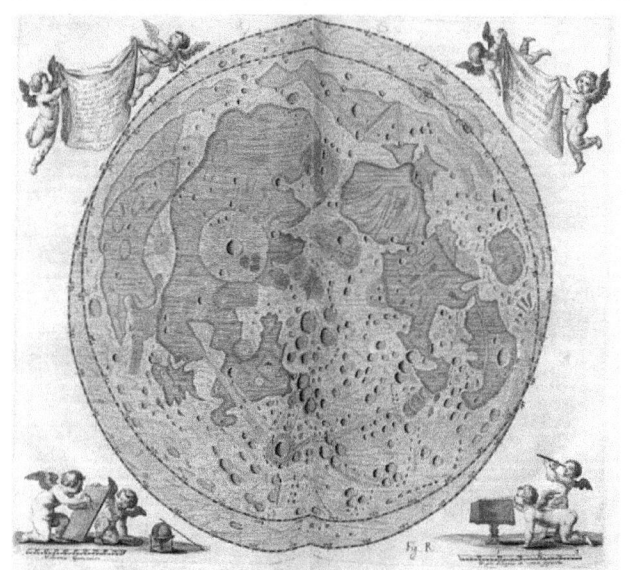

*Map of the Moon (for which element was named) by Johannes Hevelius from his 1647 "Selenographia."*
**#34, Selenium, Se**

In 1826 Antoine Jérôme Balard (French chemist) was one of two scientists to independently discover Bromine.
**#35, Bromine, Br**

Robert Bunsen (center) and Gustav Kirchhoff (left) discovered Caesium in 1860 and Rubidium in 1861.
**#37, Rubidium, Rb**
**#55, Caesium, Cs**

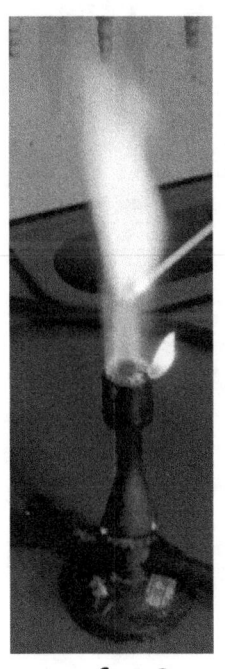

Flame test for Strontium
Strontium was discovered in Scotland in 1790.
**#38, Strontium, Sr**

One of several rare elements discovered near Ytterby, Sweden, this one in 1787.
**#39, Yttrium, Y**

*Martin Heinrich Klaproth (German chemist) discovered Uranium, Zirconium, and Cerium.*
**#40, Zirconium, Zr** (1789)
**#58, Cerium, Ce** (1803)
**#92, Uranium, U** (1789)

*Charles Hatchett (English chemist) discovered Niobium in 1801.*
**#41, Niobium, Nb**

*Climax Molybdenum Mine in Colorado, c.1924*
**#42, Molybdenum, Mo**

*Emilio Segrè, Italian physicist and one of the discoverers of Technetium and Astatine.*
**#43, Technetium, Tc** (1937)
**#85, Astatine, At** (1940)

*Gottfried Wilhelm Osann (German chemist and physicist) and Jöns Jacob Berzelius (Swedish chemist) discovered Ruthenium in 1828.*
**#44, Ruthenium, Ru**

*William Hyde Wollaston (English chemist and physicist) discovered Rhodium and Palladium in 1803.*
**#45, Rhodium, Rh**
**#46 Palladium, Pd**

*Silver mining and processing in Kutná Hora, Central Europe, 1490s.*
**#47, Silver, Ag**

*German chemists Friedrich Stromeyer (shown) and Karl Samuel Leberecht Hermann both discovered Cadmium in 1817.*
**#48, Cadmium, Cd**

German chemist Ferdinand Reich co-discovered Indium in 1863.
**#49, Indium, In**

*Washing Tin in the mines at Stanthorpe, Australia in 1872.*
**#50, Tin, Sn**

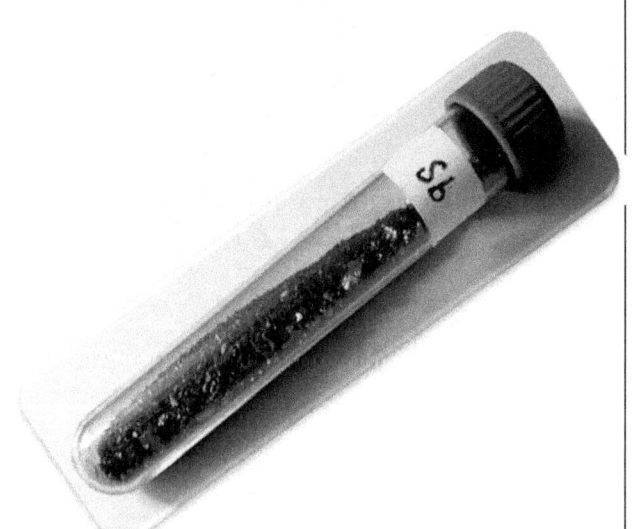

*Used in Egypt at least 5,000 years ago.*
**#51, Antimony, Sb**

*In 1782 Franz-Joseph Müller von Reichenstein (Austrian mineralogist) discovered Tellurium.*
**#52 Tellurium, Te**

*French chemist and physicist
Joseph Louis Gay-Lussac recognized
Iodine as a new element in 1811.*
**#53, Iodine, I**

*Discovered by a
Swedish chemist in 1772.*
**#56, Barium, Ba**

*Carl Gustaf Mosander,
a Swedish chemist, who discovered
Lanthanum, Erbium and Terbium.*
**#57, Lanthanum, La** (1838)
**#65, Terbium, Tb** (1842)
**#68, Erbium, Er** (1842)

*Carl Auer von Welsbach (Austrian
scientist and inventor) discovered
both elements in 1885.*
**#59, Praseodymium, Pr**
**#60, Neodymium, Nd**

*Named for Prometheus,*
*a Titan from Greek mythology, after*
*discovered in the U.S. in the 1940s.*
**#61, Promethium, Pm**

*Europium was named for Europe*
*after it was isolated*
*by a French chemist in 1901.*
**#63, Europium, Eu**

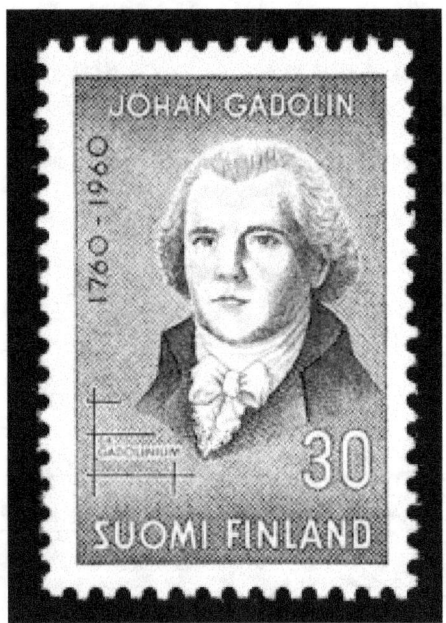

*Named for Johan Gadolin,*
*a Finnish chemist, physicist*
*and mineralogist,*
*after it was discovered in 1880.*
**#64, Gadolinium, Gd**

*Swiss chemist discovered it in 1878.*
**#67, Holmium, Ho**

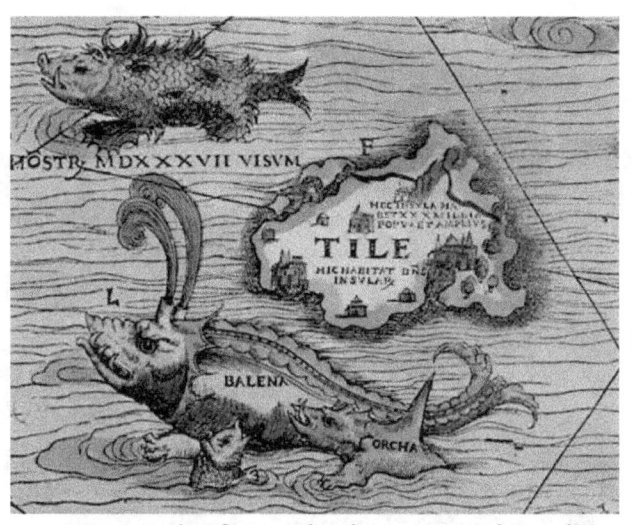

*Named after Thule, a mythical region in Scandinavia, after being isolated by a Swedish chemist in 1879.*
**#69, Thulium, Tm**

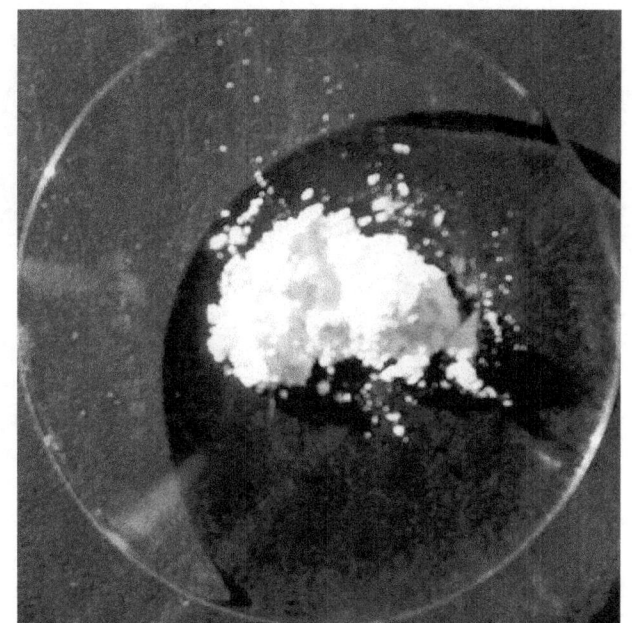

*One of several rare elements discovered near Ytterby, Sweden, this one in 1787.*
**#70, Ytterbium, Yb**

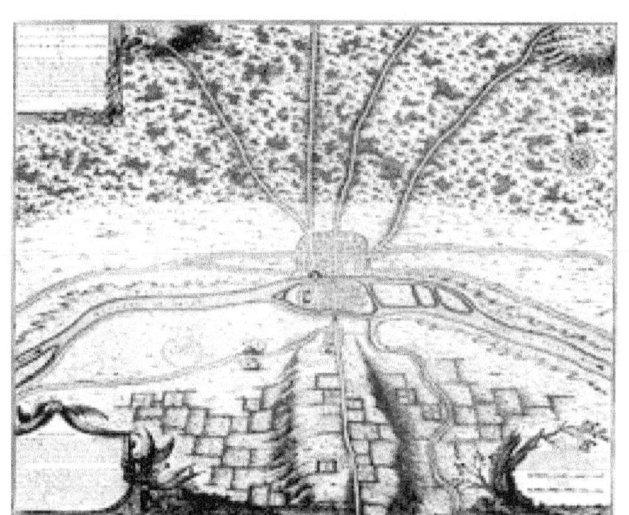

*Named for Lutetia, Latin for Paris in the Roman era, after it was discovered in 1906.*
**#71, Lutetium, Lu**

*Named after Hafnia, Latin for Copenhagen, where it was discovered in 1922.*
**#72, Hafnium, Hf**

*Anders Gustaf Ekeberg,
a Swedish chemist, who
discovered Tantalum in 1802.*
**#73, Tantalum, Ta**

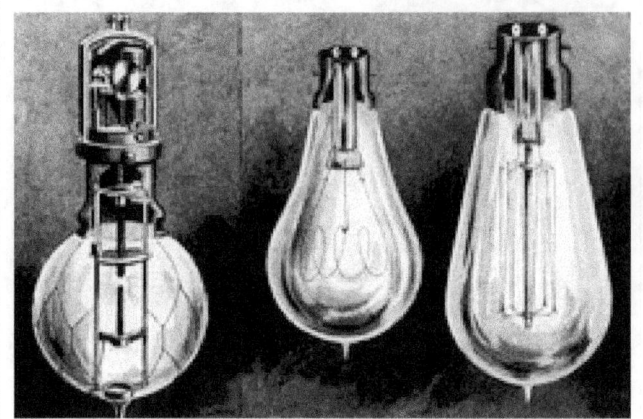

*Tungsten is often used in
incandescent light bulb filaments.
Tungsten was discovered
by a Swedish chemist.*
**#74, Tungsten, W**

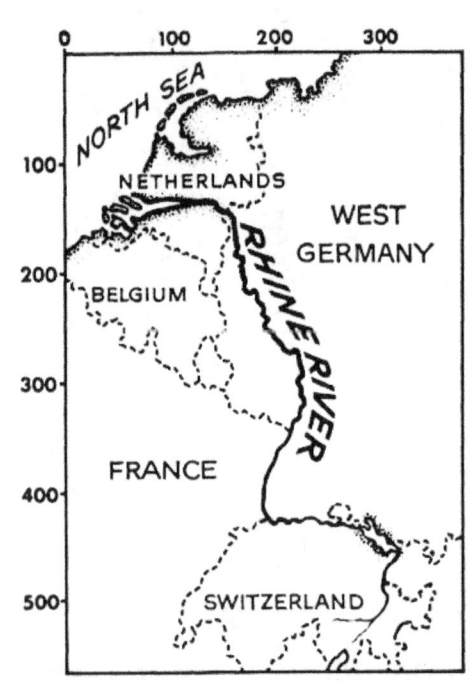

*Named after the Rhine River
after it was discovered in 1908.*
**#75, Rhenium, Re**

76     **Osmium**     Os

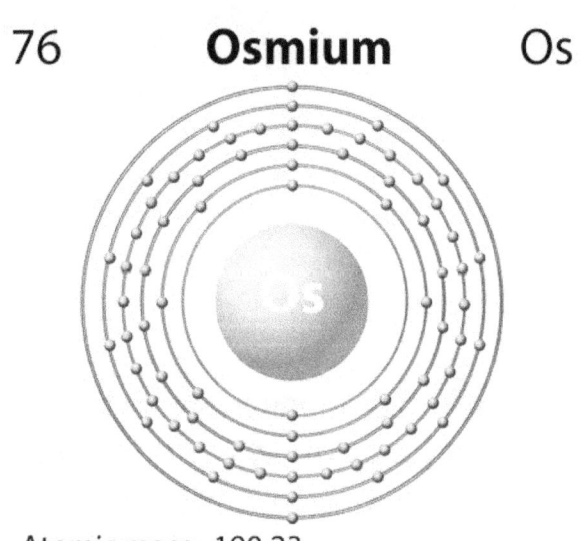

Atomic mass: 190.23
Electron configuration: 2, 8, 18, 32, 14, 2

*Discovered and isolated in
1803 by an English chemist.*
**#76, Osmium, Os**

*Named for Iris, from mythology, after it was discovered and isolated in 1803 by an English chemist.*
**#77, Iridium, Ir**

*Antonio de Ulloa (Spanish astronomer) is credited with the discovery of Platinum in 1735.*
**#78, Platinum, Pt**

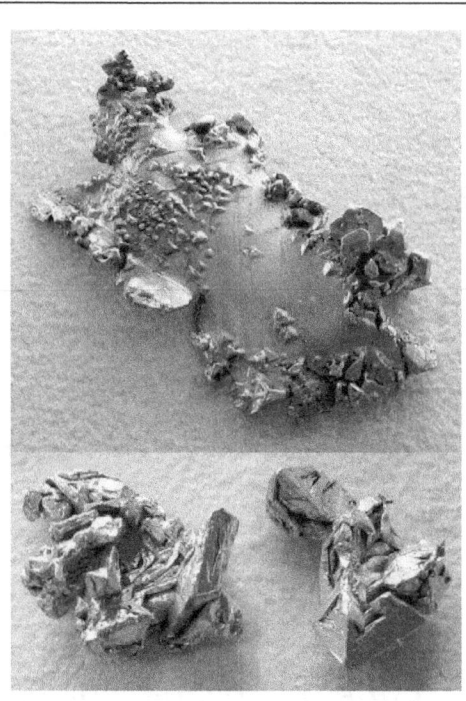

*Gold artifacts have been discovered dating back 6,000 years.*
**#79, Gold, Au**

*The Mercury Atmosphere and Surface Composition Spectrometer (MASCS) instrument aboard NASA's MESSENGER spacecraft took this photograph in 2015.*
**#80, Mercury, Hg**

*Sir William Crookes (English chemist)
discovered Thallium in 1861
(Painting by Sir Leslie Ward)*
**#81, Thallium, Tl**

*Lead mining in the upper
Mississippi River region
in the United States in 1865.*
**#82, Lead, Pb**

*Discovered in 1753
by a French chemist.*
**#83, Bismuth, Bi**

*The Curies discovered Polonium
in 1898 and named it for
Marie's home country, Poland.*
**#84, Polonium, Po**

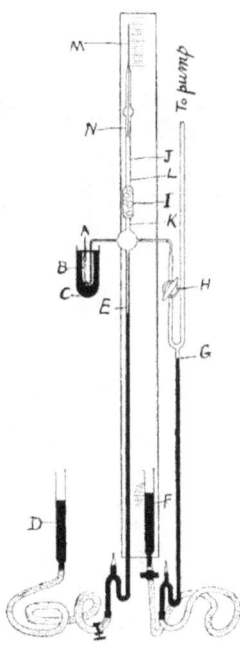

*Apparatus used by Ramsay and Whytlaw-Gray to isolate Radon in London in 1910.*
**#86, Radon, Rn**

*Marguerite Perey (a French physicist and student of Marie Currie) discovered Francium in 1939.*
**#87, Francium, Fr**

*Marie and Pierre Curie experimenting with Radium in 1898.*
**#88, Radium, Ra**

*Discovered by a German organic chemist in 1902.*
**#89, Actinium, Ac**

*Named for Thor, of Norse mythology, after being discovered by a Swedish chemist in 1829.*
**#90, Thorium, Th**

*Discovered in 1900 by an English chemist and physicist.*
**#91, Protactinium, Pa**

*Named for the planet Neptune. Discovered by American physicists in 1940.*
**#93, Neptunium, Np**

*Named for the (then) planet Pluto. Discovered by American chemists and a physicist in 1940-1941.*
**#94, Plutonium, Pu**

*These two elements were among those detected in the fallout from the Ivy Mike nuclear test the Americans conducted in 1952 (above the Pacific Ocean).*
**#95, Americium, Am**
**#100, Fermium, Fm**

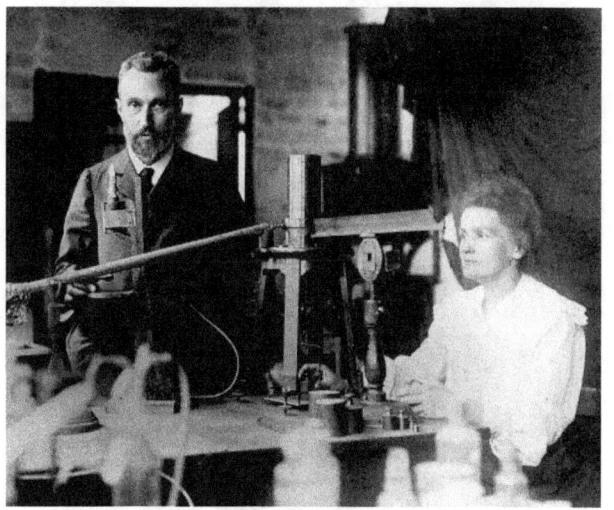

*Named for Pierre and Marie Curie when it was discovered in 1944.*
**#96, Curium, Cm**

*One of several elements discovered at the University of California in Berkeley in 1949.*
**#97, Berkelium, Bk**

*The 60-inch-diameter cyclotron used to first synthesize Californium in 1950.*
**#98, Californium, Cf**

*Named for Albert Einstein after it was discovered in 1952.*
**#99 Einsteinium, Es**

*Named for (Russian chemist) Dmitri Mendeleev, the "Father of the Periodic Table" after it was discovered in 1955.*
**#101, Mendelevium, Md**

*Named for Alfred Nobel (Swedish chemist) after it was discovered in 1966.*
**#102, Nobelium, No**

*The element was named after Ernest Lawrence (the American developer of the cyclotron) after it was discovered between 1961 and 1971.*
**#103, Lawrencium, Lr**

*Named for Ernest Rutherford
(British physicist) after it
was discovered in 1964.*
**#104, Rutherfordium, Rf**

*Named for the town in Russia
(Dubna) where it was first
produced in 1968.*
**#105, Dubnium, Db**

*Named after Glenn T. Seaborg
after it was discovered in 1964.*
**#106, Seaborgium, Sg**

*Named after Niels Bohr
after it was discovered in 1981.*
**#107, Bohrium, Bh**

*Named for the Latin name for Hesse (Germany) where it was discovered in 1984.*
**#108, Hassium, Hs**

*Named for Lise Meitner (Austrian physicist) after it was discovered in 1982.*
**#109, Meitnerium, Mt**

*Created near Darmstadt, Germany in 1994.*
**#110, Darmstadtium, Ds**

*Named for German physicist Wilhelm Conrad Röntgen after it was synthesized in 1994.*
**#111, Roentgenium, Rg**

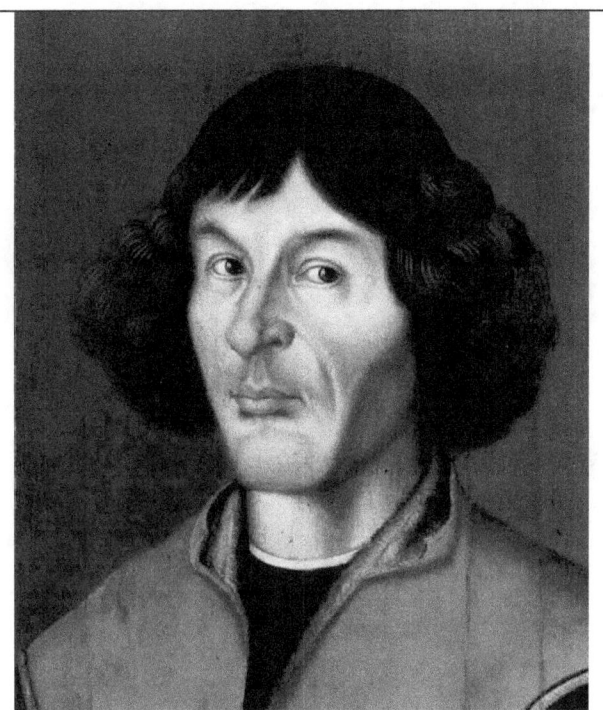

*Named for Nicolaus Copernicus after it was created in 1996.*
**#112, Copernicium, Cn**

*Temporary names for elements synthesized in Dubna, Russia in the early 2000s.*
**#113, Ununtrium, Uut**
**#115, Ununpentium, Uup**
**#117, Ununseptium, Uus**
**#118, Ununoctium, Uuo**

*Synthesized in 2000 in Dubna, Russia and at the Lawrence Livermore National Laboratory, in Livermore, CA (named for Robert Livermore).*
**#116, Livermorium, Lv**

*Named for Georgy Flyorov, founder of the Flerov Laboratory of Nuclear Reactions in Dubna, Russia, where it was discovered in 1998.*
**#114, Flerovium, Fl**

# Notes

# More Notes